DANGEROUS SNAKES

COPPERHEADS

Kelli Hicks

TABLE OF CONTENTS

A Crabtree Seedlings Book

CRABTREE
Publishing Company
www.crabtreebooks.com

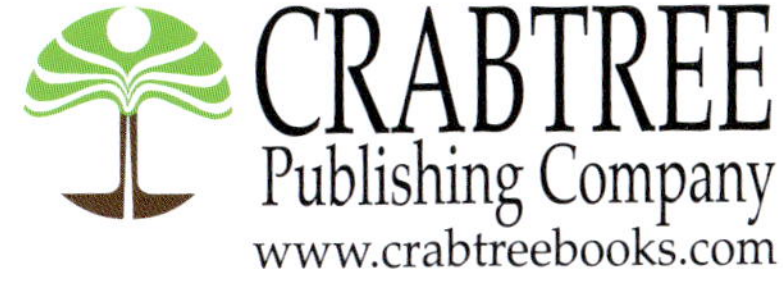

School-to-Home Support for Caregivers and Teachers

This book helps children grow by letting them practice reading. Here are a few guiding questions to help the reader with building his or her comprehension skills. Possible answers appear here in red.

Before Reading:

- What do I think this book is about?
    - *I think this book is about copperhead snakes.*
    - *I think this book will be full of cool facts about copperhead snakes.*
- What do I want to learn about this topic?
    - *I want to learn where copperhead snakes live.*
    - *I want to learn what they eat.*

During Reading:

- I wonder why...
    - *I wonder why copperhead snakes have rough skin.*
    - *I wonder why baby copperhead snakes have yellow tails.*
- What have I learned so far?
    - *I have learned that baby copperhead snakes use their yellow tail to trick their prey.*
    - *I have learned that copperheads use their fangs and venom to kill their prey.*

After Reading:

- What details did I learn about this topic?
    - *I learned that after a meal the copperhead may not eat again for more than two weeks.*
    - *I learned that they will attack when they are scared.*
- Read the book again and look for the vocabulary words.
    - *I see the word **attack** on page 20 and the word **habitats** on page 18. The other glossary words are on pages 22 and 23.*

COPPERHEADS

See that red-brown head and **rough** skin?

Look out! It is a copperhead snake.

A pattern shows on its **thick** body.

It has deadly **fangs**.

Baby copperheads are born with fangs and a yellow tail that looks like a worm. The tail tricks its **prey**.

Copperheads use their fangs and venom to kill their prey.

Copperhead fangs are hollow and work like needles to inject prey with poisonous venom.

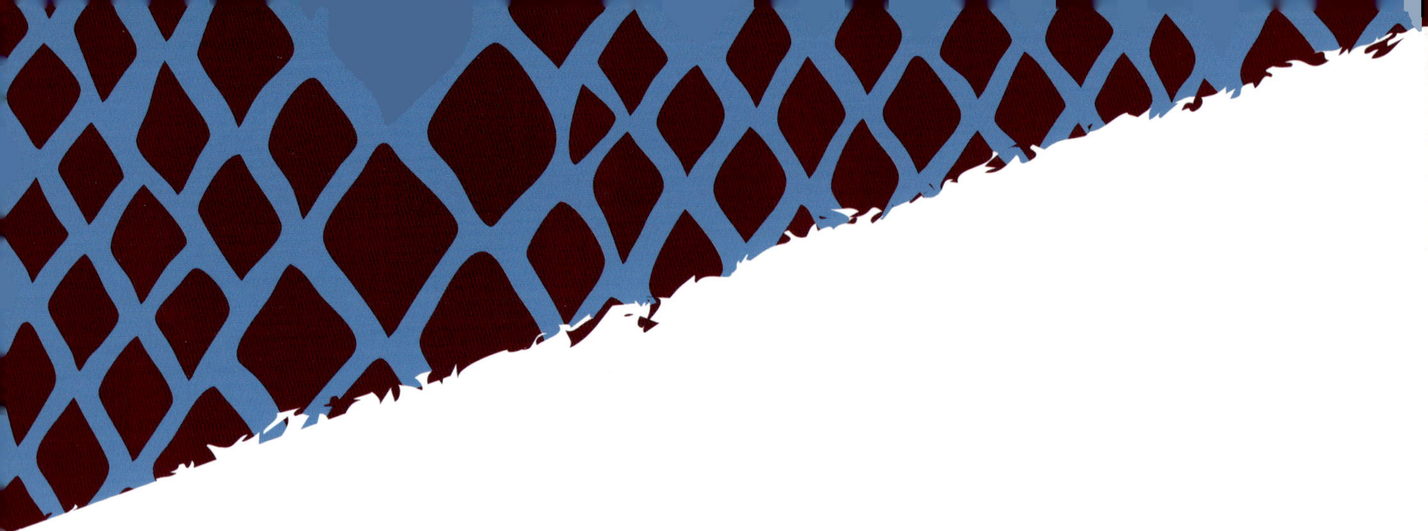

A hungry copperhead waits to strike.

It opens its mouth wide to catch its prey.

The copperhead may not eat again for more than two weeks.

Copperhead snakes can live in different **habitats**.

Copperheads live in forests, deserts, and even in swampy places.

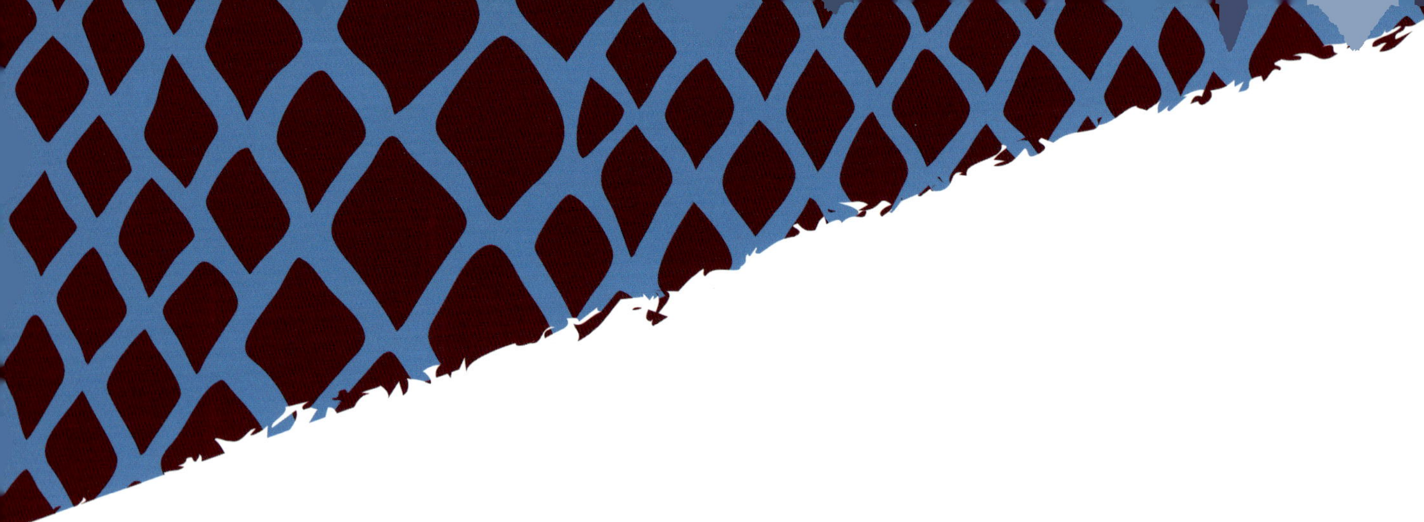

Be careful! The snake will **attack** when scared.

Glossary

attack (uh-TAK): To attack means to try to hurt someone or something.

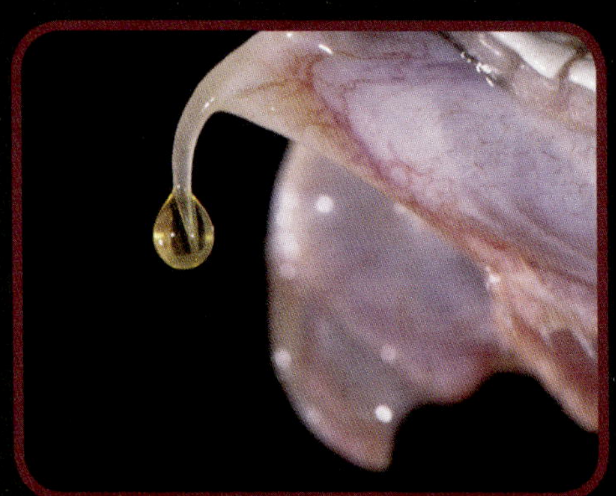

fangs (FANGZ): Fangs are sharp, long teeth.

habitats (HAB-uh-tats): Habitats are the places animals and plants naturally live.

prey (PRAY): Prey is an animal that is hunted by another animal for food.

rough (RUHF): Rough surfaces have bumps. They are not smooth.

thick (THIK): Thick means wide or deep, not thin.

Index

About the Author

Kelli Hicks

Kelli Hicks loves to learn about science and nature, including dangerous snakes. She prefers to read about snakes rather than meet them in person. She lives in Tampa with her husband and two children and her dog, Emma June.

Websites

https://www.livescience.com/43641-copperhead-snake.html
https://www.nationalgeographic.com/animals/reptiles/c/copperhead-snakes/

Written by: Kelli Hicks
Designed by: Jennifer Dydyk
Edited by: Tracy Nelson Maurer

Photographs:
mask for snakeskin graphic on cover and pages © shutterstock.com/Merydolla; yellow triangle with snake graphic © Top Vector Studio/Shutterstock; Cover photo: © shutterstock.com/ Wildvet. Page 3: ©shutterstock.com/Matt Jeppson. Page 5: ©shutterstock.com/Dennis W Donohue. Page 6: ©shutterstock.com/Dennis W Donohue. Page 7: ©stock.com/amwu. Page 8: ©Laura Ballard | Dreamstime.com. Page 9: ©shutterstock.com/Breck P. Kent. Page 11 ©shutterstock.com/Suzanna Ruby. Page 13: ©shutterstock.com/Matt Jeppson. Page 15: ©shutterstock.com/Dennis W Donohue. Page 17: ©istock.com/makasana. Page 19: ©shutterstock.com/lev radin. Page 21: ©William Wise | Dreamstime.com. Page 22 fang: © shutterstock.com/ Joe McDonald. Page 23: ©istock.com/CreativeNature_nl.

Library and Archives Canada Cataloguing in Publication

Title: Copperheads / Kelli Hicks.
Names: Hicks, Kelli L., author.
Description: Series statement: Dangerous snakes | "A Crabtree seedlings book". | Includes index.
Identifiers: Canadiana (print) 20210202203 | Canadiana (ebook) 20210202211 | ISBN 9781427162342 (hardcover) | ISBN 9781427159090 (softcover) | ISBN 9781427159106 (HTML) | ISBN 9781427159113 (EPUB) | ISBN 9781427159120 (read-along ebook)
Subjects: LCSH: Copperhead—Juvenile literature.
Classification: LCC QL666.O69 H53 2022 | DDC j597.96/3—dc23

Library of Congress Cataloging-in-Publication Data

Names: Hicks, Kelli L., author.
Title: Copperheads / Kelli Hicks.
Description: New York : Crabtree Publishing, [2022] | Series: Dangerous snakes- a Crabtree seedlings book | Includes index.
Identifiers: LCCN 2021018643 (print) | LCCN 2021018644 (ebook) | ISBN 9781427162342 (hardcover) | ISBN 9781427159090 (paperback) | ISBN 9781427159106 (ebook) | ISBN 9781427159113 (epub) | ISBN 9781427159120
Subjects: LCSH: Copperhead--Juvenile literature.
Classification: LCC QL666.O69 H53 2022 (print) | LCC QL666.O69 (ebook) | DDC 597.96/3--dc23
LC record available at https://lccn.loc.gov/2021018643
LC ebook record available at https://lccn.loc.gov/2021018644

Crabtree Publishing Company

www.crabtreebooks.com 1-800-387-7650

Printed in the U.S.A./062021/CG20210401

In Canada: We acknowledge the financial support of the Government of Canada through the Canada Book Fund for our publishing activities.

Published in the United States
Crabtree Publishing
347 Fifth Avenue, Suite 1402-145
New York, NY, 10016

Published in Canada
Crabtree Publishing
616 Welland Ave.
St. Catharines, Ontario L2M 5V6